# HANDBOOK
## OF
# SHOE FACTORY ENGINEERING

# A History of Shoemaking

Shoemaking, at its simplest, is the process of making footwear. Whilst the art has now been largely superseded by mass-volume industrial production, for most of history, making shoes was an individual, artisanal affair. 'Shoemakers' or 'cordwainers' (cobblers being those who repair shoes) produce a range of footwear items, including shoes, boots, sandals, clogs and moccasins – from a vast array of materials.

When people started wearing shoes, there were only three main types: open sandals, covered sandals and clog-like footwear. The most basic foot protection, used since ancient times in the Mediterranean area, was the sandal, which consisted of a protective sole, attached to the foot with leather thongs. Similar footwear worn in the Far East was made from plaited grass or palm fronds. In climates that required a full foot covering, a single piece of untanned hide was laced with a thong, providing full protection for the foot, thus forming a complete covering. These were the main two types of footwear, produced all over the globe. The production of wooden shoes was mainly limited to medieval Europe however – made from a single piece of wood, roughly shaped to fit the foot.

A variant of this early European shoe was the clog, which were wooden soles to which a leather upper was attached. The sole and heel were generally made from one piece of maple or ash two inches thick, and a little longer and broader than the desired size of shoe. The outer side of the sole and heel was fashioned with a long chisel-edged implement, called the clogger's knife or stock; while a second implement, called the groover, made a groove around the side of the sole. With the use of a 'hollower', the inner sole's contours were adapted to the shape of the foot. In even colder climates, such designs were adapted with furs wrapped around the feet, and then sandals wrapped over them. The Romans used such footwear to great effect whilst fighting in Northern Europe, and the native Indians developed similar variants with their ubiquitous moccasin.

By the 1600s, leather shoes came in two main types. 'Turn shoes' consisted of one thin flexible sole, which was sewed to the upper while outside in and turned over when completed. This type was used for making slippers and similar shoes. The second type united the upper with an insole, which was subsequently attached to an out-sole with a raised heel. This was the main variety, and was used for most footwear, including standard shoes and riding boots.

Shoemaking became more commercialized in the mid-eighteenth century, as it expanded as a cottage industry. Large warehouses began to stock footwear made by many small manufacturers from the area. Until the nineteenth century, shoemaking was largely a traditional handicraft, but by the century's end, the process had been almost completely mechanized, with production occurring in large factories. Despite the obvious economic gains of mass-production, the factory system produced shoes without the individual differentiation that the traditional shoemaker was able to provide.

The first steps towards mechanisation were taken during the Napoleonic Wars by the English engineer, Marc Brunel. He developed machinery for the mass-production of boots for the soldiers of the British Army. In 1812 he devised a scheme for making nailed-boot-making machinery that automatically fastened soles to uppers by means of metallic pins or nails. With the support of the Duke of York, the shoes were manufactured, and, due to their strength, cheapness, and durability, were introduced for the use of the army. In the same year, the use of screws and staples was patented by Richard Woodman. However, when the war ended in 1815, manual labour became much cheaper again, and the demand for military equipment subsided. As a consequence, Brunel's system was no longer profitable and it soon ceased business.

Similar exigencies at the time of the Crimean War stimulated a renewed interest in methods of mechanization and mass-production, which proved longer lasting. A shoemaker in Leicester, Tomas Crick, patented the design for a riveting machine in 1853. He also introduced the use of steam-powered rolling-machines for hardening leather and cutting-machines, in the mid-1850s. Another important factor in shoemaking's mechanization, was the introduction of the sewing machine in 1846 – a development which revolutionised so many aspects of clothes, footwear and domestic production.

By the late 1850s, the industry was beginning to shift towards the modern factory, mainly in the US and areas of England. A shoe stitching machine was invented by the American Lyman Blake in 1856 and perfected by 1864.

Entering in to partnership with Gordon McKay, his device became known as the McKay stitching machine and was quickly adopted by manufacturers throughout New England. As bottlenecks opened up in the production line due to these innovations, more and more of the manufacturing stages, such as pegging and finishing, became automated. By the 1890s, the process of mechanisation was largely complete.

Traditional shoemakers still exist today, especially in poorer parts of the world, and do continue to create custom shoes. In more economically developed countries however, it is a dying craft. Despite this, the shoemaking profession makes a number of appearances in popular culture, such as in stories about shoemaker's elves (written by the Brothers Grimm in 1806), and the old proverb that 'the shoemaker's children go barefoot.' Chefs and cooks sometimes use the term 'shoemaker' as an insult to others who have prepared sub-standard food, possibly by overcooking, implying that the chef in question has made his or her food as tough as shoe leather or hard leather shoe soles. Similarly, reflecting the trade's humble beginnings, to 'cobble' can mean not only to make or mend shoes, but 'to put together clumsily; or, to bungle.'

As is evident from this short introduction, 'shoemaking' has a long and varied history, starting from a simple means of providing basic respite from the elements, to a fully mechanised and modern, global trade. It is able to provide a fascinating insight not only into fashion, but society, culture and climate more generally. We hope the reader enjoys this book.

## INTRODUCTION

The material and data contained within this little book are the result of many years' investigation and experience in the designing and equipping of modern factories—particularly as applied to the manufacture of shoes.

The remarkable economies and increased efficiency to be obtained by conducting manufacturing operations in buildings especially designed to meet the particular requirements of the industry with full consideration paid to the modern development of those important elements which have to do with light, heat, power, equipment and interior organization, have but recently attracted the attention of shoe manufacturers.

The remarkably rapid growth of the shoe manufacturing industry during the past fifteen years has afforded shoe manufacturers but slight opportunity for investigation in this important field, the increased demand for more space being ordinarily met by building additions to old plants. In a constantly increasing number of instances, however, this expedient has reached its logical limit and brought those concerned in the factory management face to face with the building problem.

Anticipating this tendency in the evolution of the industry, the experts of the Department of Agencies have from its inception been closely in touch with every phase of its development and plans for many of the factories of most modern construction have been prepared by them.

The services and experience of these experts are at the disposal of the Company's patrons and the constantly increasing demand for data in this connection has brought this little book into being.

With a full realization of the proportions of the problem and the impossibility of treating in more than a general way at this time so vast a subject, the important information it contains is submitted as a basis for more detailed and thoughtful consideration.

# AN IDEAL SHOE FACTORY

## THE BUILDING

If no disturbing conditions or requirements intervened and the manufacturer was able to accept the following specifications (detailed explanation of which appears on another page), he would have a factory, ideal in construction and arrangement. It is seldom indeed that this ideal condition can be found, but the details should be modified only as necessity requires. The ideal factory would conform to this schedule: —

Shape: Rectangular
Width: 40 to 50 feet inside
Length: As per capacity, up to 1,000 feet
Number of Floors:
   3 Floors and Basement
   Top Floor — Cutting and Upper Fitting
   Second Floor — Lasting and Making
   Street Floor — Finishing, Dressing and Packing
   Basement — Stock Fitting, Sole Leather Cutting and Storage
Altitude of Floors:
   9 feet, 6 inches (minimum.)
   11 feet, 6 inches (ideal)
Floor Strain:
   150 pounds per square foot
Tower: One to every 300 or 400 feet. Located in the centre of side, 40 feet by 40 feet
   Its function:
      Elevators, stairways, toilets, coatrooms, etc.
Windows:
   As many as possible
   Sawtooth roof
Construction material:
   Relative as to locality

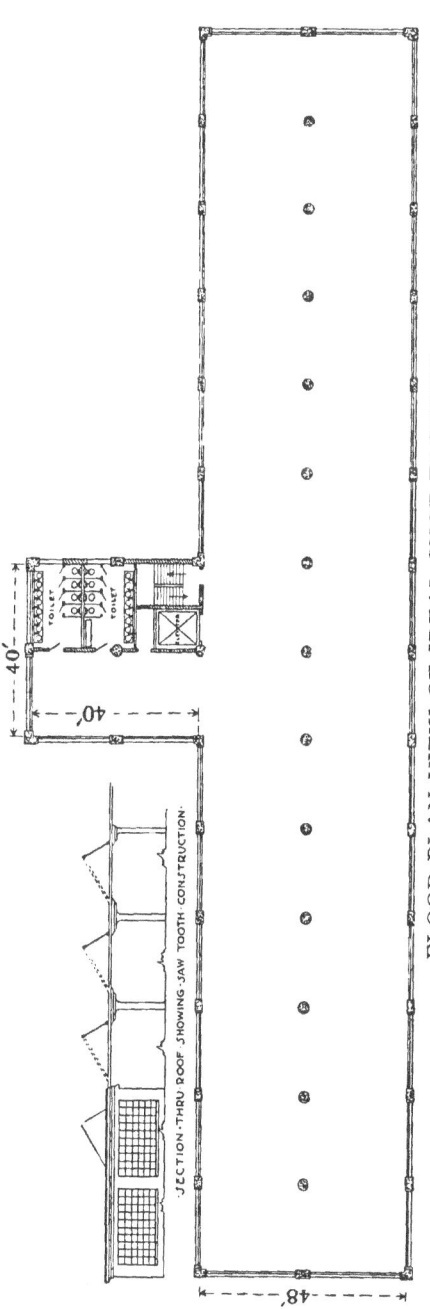

FLOOR PLAN VIEW OF IDEAL SHOE FACTORY

One row of posts. Tower of ample size, leaving the manufacturing space straightaway for shoemaking. With a sawtooth roof the steady north light makes the central portion of top floor as light as the sides.

Cost; approximate.
- Concrete — cost $1.25 square foot
- Brick — cost $1.00 (150 pounds) square foot
- Wood — cost $ .75 to $ .90 (150 pounds) square foot

Approximate number of feet necessary per pair of shoes:
- Welts — 15 square feet
- Turns — 15 square feet
- McKay — 10 to 12 square feet
- Standard Screw and Pegs — 10 to 12 square feet
- Stitchdowns — 8 to 10 square feet

Offices:
- Location:
  - Separate Building
  - End of building, first floor

Provision for Employees' Welfare:
- Rest room, restaurant, recreation room, hospital, etc.

Storage for Material:
- Separate building
- Small supply of finished and raw stock in factory

Humidifying Room:
- Important for the proper conditioning of sole leather, waxes, threads, counters, etc.
- Size of room as per conditions.

---

In the consideration of a shoe manufacturing plant, the first problem which presents itself is: "What shape shall it be?"

**Shape and Size**
"What is the best?" Years of experience have proven that a rectangular shaped building, from forty to fifty feet wide with the length up to the limit of one thousand feet, works out to the best advantage. Fifty feet in width is practically the limit for a factory where only one row of posts is desired for support in the middle, and fifty feet is preferable to forty, due to the fact that it allows ample room for the storage of shoes awaiting

ENDICOTT JOHNSON CO. "SCOUT" FACTORY, JOHNSON CITY, N. Y., 620 FEET LONG, SHOWING POSITION OF TOWERS

The building is of interest on account of its length and width, simplicity of design, steel super-structure with brick facing and the Dust Collecting System. Motor and Fans being on outside wall of building.

their operation and gives plenty of aisle space for the incoming and outgoing racks. As it is well known that the central portion of a shoe factory is not light enough for machine operation, the rectangular shaped building should need no further defence.

Three stories and a basement is the best height for the building. If we add another floor, or floors, the strain **Height** becomes so great that the cost of construction increases materially. The best arrangement for the operation in three stories, due to the fact that the natural sequence of operations follow logically and there is, therefore, less handling of the shoes, is as follows:

    Top Floor              Street Floor
       Cutting                Finishing
       Upper Fitting        Dressing
                                 Packing
    Second Floor         Offices
       Lasting
       Making                Basement
                                 Stock Fitting
                                 Sole Leather Cutting
                                 Storage

The altitude of the floors should not be less than nine feet, six inches; and eleven feet, six inches is ideal. The latter **Altitude** allows for better ventilation and light, and gives ample head room under the pulleys, belts, etc., for the operatives.

The tower in connection with any manufacturing plant, especially the shoe plant, is modern and a feature that, if **Tower** properly located and designed, is well nigh indispensable. It should be on the outside, in the centre of the building if under five hundred feet in length, or there should be a tower for every three or four hundred feet. Forty feet square is the ideal size. By the use of this, or these towers for the location of elevators, stairways, toilets, coat rooms, etc., the main factory is left entirely for the purpose for which it was intended; viz — the manufacture of shoes.

When designing a factory as much window space as possible should be provided. There is no one who will argue that you
**Windows** can have too much light. The roof should be of the sawtooth construction, admitting the north light in order that the central portion of the room will be as well lighted as the sides. This allows for the best possible arrangement of the benches in the Upper Fitting Room.

There are three choices as to what is best to use for material in the construction of a shoe factory. Concrete is fire-proof,
**Building Material** requires less upkeep and admits of the lowest rate of insurance, but costs more to construct in the beginning. It is possible with concrete to allow room for more window space than with any other material.

Brick, with the beam and girder construction is the second choice, at a little less figure than the concrete.

Wood is the last choice. It costs two-thirds as much as

LUPTON STEEL SASH FOR SIDE WALLS

POND CONTINUOUS SASH IN SAWTOOTH ROOF
CONSTRUCTION

the concrete, but would require a high rate of insurance and constant repairs.

The location of the factory would influence materially the decision as to which of the three above types would be most desirable.

The prices of material will vary greatly in different sections of the country.

The offices, if possible, in a large manufacturing plant should be in a separate building, in order that they may be away from **Offices** the noise and leave additional room for the manufacturing. If, however, the resources of the manufacturer are limited, the best place for the offices is on the street floor at one end of the building.

Most employers of labor in these modern times plan for their employees' welfare by the installation of rest rooms,

**Employees' Welfare**

hospitals, restaurants and recreation rooms, and in designing a new factory it would seem a mistake to leave this feature without consideration. It is hard to say what is the best location. If it were a city factory, a portion of the top floor would probably be the best, due to the fact that there would be more light and

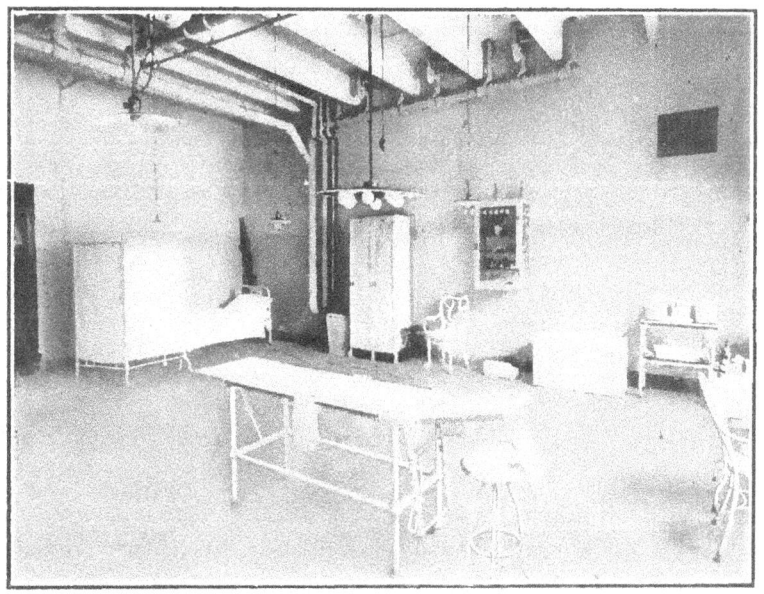

HOSPITAL AT PLANT OF U. S. M. C. BEVERLY, MASS.

air, but if in the country perhaps the conveniences of a properly arranged basement might be more desirable.

The manufacturer of shoes should not use his manufacturing plant, for which he charges himself a fair rental rate, for the

**Storage of Material**

use of storage of material, and never should he carry in his plant more than a week's supply of raw material or finished goods. If it is necessary to stock material in large quantities, either finished or raw, a separate building or a rented loft should be arranged for at a much less cost than which he charges against himself for manufacturing.

ONE OF THE LOCKER AND WASH ROOMS AT
U. S. M. CO. FACTORY

**Humidifying Room**

Provision for a suitable sized-room for the proper conditioning of leather should be arranged for in which a moistening or humidifying machine can be placed. This room should be located near the assembling department, and the soles, counters, etc., sent from there to the various departments as per the requirements.

## DESCRIPTIVE DETAILS OF TEMPERING ROOM

It is beneficial in the conditioning of leather: — Welting, leather for soles, counter blanks, etc. — when the temper is once reached, to retain its mulled condition, without having

HUMIDIFYING ROOM IN FACTORY OF LUND & SWEET,
AUBURN, ME.

recourse to wetting again, and to this end the tempering or humidifying room has been developed.

Any room of proper dimensions, available, will do, but better, a room may be built in that portion of the factory next to the assembling department, wood sheathing from the floor to the ceiling, and of such size to give the cubical contents necessary for the storage of a sufficient number of dozens of pairs of outer soles and other parts of the shoe which require tempering. This room should have proper window lights to permit the free circulation of air, and the entering of a proper amount of sunlight, and by means of the Humidor therein installed, the air may be kept at a certain percentage of humidity without depriving the leather of sunlight and fresh air, necessary to prevent "blooming" or "spuming" which commonly occurs when leather is stored in a dark room that is more or less air-tight, causing of course, the quality to deteriorate.

The tempering room is arranged in a proper manner for the storing of leather by means of shelves of pre-determined width and height, and made of dowel-pin stock to permit of the proper circulation of air. This method of tempering assures splendid results, and the leather will be maintained at a proper working condition with a minimum effort, and a greater increase of value from the shoe-making standpoint.

A room 20 by 14 by 12 feet, contains 3,360 cubic feet; — the Humidifier being suspended from the ceiling, dropping 30 inches below the same, is operated by a small fan motor Electricity may be taken from any convenient electric light socket, and controlled by three-speed switch, so that the conditions may be readily controlled and varied as desired. The ½-inch water supply is taken from any convenient source and supplies the water jet, which vaporizes the water absorbed by the air passing thru the Humidifier, the maximum capacity of which is 26,000 cubic feet of air per hour, with a consumption of 35 gallons of water per hour (more or less); — at a pressure of 50 pounds — 1 gallon per hour is absorbed by the air. The volume of water may be increased by maintaining a higher water pressure.

The water control is independent of the air circulation, and can be shut off entirely independent of the air current, which is desirable to maintain at all times, as it produces a uniform effect in the room, all the air of the room passing thru the Humidor, and being washed and properly humidified, thereby assuring a uniform circulation of the air and condition of the stock. A normal temperature of the room is only required, the air in the room being circulated seven times per hour.

# AN IDEAL SHOE FACTORY

## POWER AND LIGHT

If the manufacturer is not confronted by any unusual conditions and is free to choose, the facts presented in the following schedule regarding these two very important elements should receive most careful consideration: —

Requirements, (approximated):
  20-25 pair Welts per H. P.
  20-25 pair Turns per H. P.
  30-40 pair McKays per H. P.
  30-40 pair Nailed and pegged per H. P.
  40-50 pair Stitchdowns per H. P.

Electric Motors:
  550 volts, 3 phase, 60 cycle
  Alternating Current
  1800 R. P. M., all departments except Upper Fitting Room
  1200 R. P. M. Upper Fitting Room
  720 R. P. M. Exhaust Fan
  Oil Switches for 5 H. P. Motors or less

Turbine, (Curtis Type):
  For 100 H. P. requirements and over
  For small requirements
    Reciprocating steam engine and dynamo
    Gas engine and dynamo
    Purchased current

Cost of making power, light and heat
  1½ to 2 cents per K. W.

Cost of buying power
  Dependent on amount purchased from 1½-3½ cents per K. W.
  For lighting alone in most districts the rate is three times greater than for power.

Lighting:
: General Illumination
:: Suitably powered lamps and shades at altitude of not less than 9 feet
: Specific Illumination
:: As conditions warrant

Millwrighting:
: Standard size of shafting 1-$\frac{15}{16}$ inches in diameter and 60 foot lengths for all departments except upper fitting room; 1-$\frac{3}{16}$ inches diameter for Upper Fitting Room.
: Sells Roller Bearings for shafting, reducing frictional load to a minimum.
: Hangers, 10 to 12 inch drop
: Speed of shafting:
:: 300 R. P. M. except in Upper Fitting Room
:: 350 R. P. M. Upper Fitting Room
: Steel Pressed Pulleys are most satisfactory
: Transmission for Upper Fitting Room:
:: Link Silent Chain Belt
:: Spartan "V" Belts

Dust Collecting System:
: Slow speed, high pressure is most economical
: Location: outside of building if possible
: 20-22 Gauge Piping
: Cost of Installation:
:: $30-$40 for every dust making machine, exclusive of motors

---

The kind of a Power Equipment that should be put in a shoe factory is a most important factor to consider. The generally accepted method of installation is with Short Drives and Electric Motors. The old-fashioned method of Long Line Drives belted from floor to floor and then to the engine is no longer modern or efficient.

On using the old Long Drives starting with large shafting at the engine end and gradually working down to the small

3 H. P. MOTOR REPLACING "MULE STAND DRIVE"
SPRINGFIELD SHOE WORKS

5 H. P. MOTOR DRIVING 12 HAND METHOD LASTING
MACHINES AND 4 REX PULLING-OVER MACHINES

**Old Method** "pipe-stem" shafting, an enormous amount of power is wasted in frictional load for turning the shafting alone, and waste of power means waste of coal and coal is expensive.

By the use of electric motors and short drives, we have the most flexible kind of an outfit, and if increases of output are contemplated this flexibility works out very simply by the addition of new or larger motors, while with the steam-driven plant an increase in the output requires a large investment for the new requirements.

**New Method**

**Type of Motors** The Alternating Current type of motor of 550 volts, 3-phase and 60-cycle, has proven the most satisfactory.

The 550-volt circuit requires less cost to install on account of the small size wire which can be used, and this high voltage should not be more dangerous than a lower one, provided that the necessary safety appliances, such as oil switches and up-to-date methods of wiring, are used; and as this high voltage current will demand such devices, this voltage really proves less dangerous than a lower one.

**High Voltage**

Alternating Current Motors are most satisfactory, due to the intermittent factor of Shoe Machinery, and motors of this type can be depended upon to take care of instantaneous overloads where Direct Current Motors would be soon damaged by such abuse.

**Alternating Current**

A most important question next comes up as to whether it is feasible to buy or manufacture electric current. Generally speaking, unless one has a requirement of 100 horse power, it is best to buy, and for a larger requirement to install a power plant of his own, unless the Central Power Station rate is very low indeed. This is a case for figuring. It should, however, be borne in mind that a factory must be heated, and to heat a building and furnish certain machines with steam, a boiler must be operated and at high pressure, for the Goodyear Wax Thread Machines require a pressure of from thirty-five

**Cost of Power Plant**

100 K. W. ALTERNATING CURRENT CURTIS STEAM TURBINE

750 K. W. TURBINE

to forty pounds; and when one has such a boiler or boilers they should be worked to their capacity in order to show true economy.

The Curtis type of turbine engine is a most satisfactory installation for a 100 K. W. requirement or over. It is simple and compact in design, and is economical as well as being modern in construction. The Bleeder type permits of the taking of the exhaust steam after its first stage in sufficient quantities to insure the proper heating of the building.

**Turbine**

There are several types of condensing outfits on the market for use in connection with the turbine, and the engineer in charge should be best informed as to which type would be most suitable in each instance.

**Condensers**

For smaller requirements than 100 K. W. a direct connected generator with a reciprocating steam or gas engine might best answer the needs of a manufacturer.

**Other Power Units**

DIRECT CONNECTED GENERATOR SET AND
SWITCH BOARD

**Cost of Power**  The cost of making or buying current should not exceed two or two and one-half cents per K. W., except when purchased in small quantities.

**Lighting**  The artificial lighting of a factory is a feature which has been the subject of considerable study during the past few years, and experience has proven that, under most conditions, it is possible to provide sufficient light by means of general illumination properly located rather than by specific lights for each machine.

**Millwrighting**  Shafting $1\frac{15}{16}$ inches in diameter throughout the plant, with the exception of the Upper Fitting Room, seems to be rather an ideal condition. This should be put up in units of 60 feet, and located 18 inches from the wall and have a span of 8 feet between the boxes, 10 feet being the limit for any span which we should recommend.

**Bearings**  Sells Roller Bearings are ideal for the boxes, as this type of bearing reduces the frictional load to about 25 per cent, whereas with Babbitted Boxes the frictional load, under the best installations, seldom betters 40 per cent. A 10 or 12 inch drop to the hangers should warrant a satisfactory condition.

**Speed of Shafting**  The shafting should run 300 revolutions per minute, except in the Upper Fitting Room, for at this speed the Driving Pulleys average a convenient size, and the belt lengths to the various machines are not excessive.

**Pulleys**  Steel pressed pulleys are recommended, and, although the first cost is a fraction more, the satisfaction and wear are worth the difference.

**Millwrighting for Upper Fitting Room**  The Upper Fitting Room Millwrighting is special owing to the type of bench most universally used, and to the high speed of the upper sewing machines. The shafting should be $1\frac{3}{16}$ inches in diameter, and should run at 350 revolutions per minute. The shafting should be connected with motors by a silent chain belt to insure a uniform speed under all conditions. The Spartan "V" Belt can be conveniently used in connecting two or more benches where a separate motor for each bench is not practical.

A slow speed, high pressure, dust collecting system is an economical method of taking care of the dust problem. The collector should be located convenient to the power house, and the piping should, if possible, be on the outside of the building. This seems the logical location, owing to the fact that when it is put on the inside of the building it takes up too much room. In placing it on the outside it is necessary to use a somewhat heavier gauge piping, and this should, of course, be kept painted in order to make it weather-proof. A rough way of figuring the cost of installing a dust collecting system is to figure from $30 to $40 for each dust making machine; this figure does not include the cost of the motors for driving the fans.

**Dust Collecting**

LINK BELT INSTALLATION IN UPPER FITTING ROOM

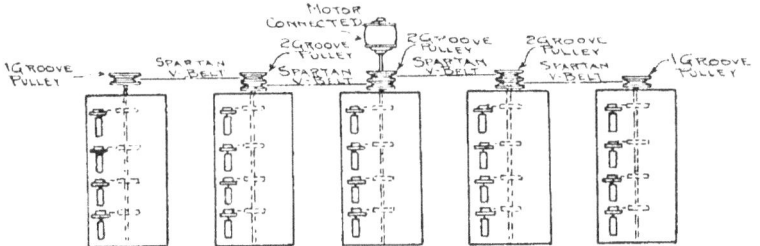

PLAN OF UPPER FITTING TABLES CONNECTED WITH SPARTAN OR LINK "V" BELTS

OVERHEAD EXHAUST SYSTEM

TWO EXHAUST FANS ON OUTSIDE WALL,
ENDICOTT, JOHNSON & COMPANY

"SELLS" ROLLER BEARING BOXES WITH SINGLE
ROLLER STRUCTURE FOR LINE SHAFTS AND
COUNTER SHAFTS

In operating a manufacturing plant, there is no one item of expense of more importance than that of the "friction load" of the power-transmitting machinery.

That this "friction load" can be reduced from 25% to 50%, depending upon load, speed, etc., has during the past seven years been proven beyond any possibility of doubt the only question remaining to be determined being the type of Roller or Ball Bearing it is desirable to use.

A series of tests and actual use, for a period of six or seven years, has demonstrated that a Roller Bearing is better than a Ball Bearing for transmitting power in a mill driven by steam, water or electricity.

CARTON MAKING EQUIPMENT

## THE AUTOBOX CARTON SYSTEM

The Autobox Carton System is particularly valuable to shoe factories because it assures quality, convenience and economy.

The box is made from a single piece of boxboard, the corners of which are retained, and folded across inside the ends, thereby doubling the ends and making them the strongest and stiffest part of the carton.

Shoe manufacturers give away a carton with every pair of shoes sold.

Carton cost is, just as legitimately as the leather cost, a part of the cost of making and marketing shoes. Therefore any saving in carton cost is just as vital; and in most shops the saving is easier to make.

The manufacture of acceptable shoe cartons is not a single operation employing a single machine, but a progressive operation involving several processes.

The first operation is the preparation of the box blank, which is cut out and scored at one operation on a cutting and creasing press. Next follow the gluing, folding and setting up, which are performed automatically at one operation by the Autobox, and which completes the shell of the box.

The blank for the lid is cut and scored by the same method and on the same machine as the box blank — the die being changed, of course. The lid blank is folded by hand or by machine as may be desirable, according to the output, and is "stayed" at the corners with Manila or Kraft Stock, on a Staying Machine, thus completing the shell of the lid.

Both box and lid shells are then "stripped" with glazed or colored paper, the operation being performed on "Stripping Machines," of which there are several types, adapted to the different requirements of different shops; and the lid is "topped" on a "Topping Machine," usually with paper the same as the strip. These operations cover the edges of the blank and give a finished carton.

Machines for mixing the adhesives, for slitting the covering paper and for baling the waste are usually made a part of the carton plant.

The power required for the operation of the plant is comparatively slight, amounting to not more than three or four horsepower in a 10,000 pair plant. Live steam from a half-inch pipe is desirable for heating and mixing adhesives; but if this is not conveniently obtainable, cold adhesives may be used at but little difference in cost.

The employees, excepting for blanking and for the Autobox, are usually women, and most of the operations are paid for on a piece basis.

# HANDY REFERENCE FIGURES

Steam requirements in pounds for heating:
  Goodyear Stitchers .................. 35-40 lbs.
  Goodyear Welters ................... 35-40 lbs.
  McKay & Richardson Sewers ......... 40 lbs.
  Bottom Driers ..................... 35 lbs.

Cost of heating a factory by steam in a climate like New England: approximated

  3 cents to 4 cents per square foot of floor space per season.

Boiler cost:
  200 pounds pressure Babcock & Wilcox Co., about $16.00 per horse power
  International Engineering Works, Ltd., about $11.00 to $13.00 per horse power.

These prices do not include piping, but do include delivery and setting

Motor cost:
  From $16.50 to $20.00 per horse power

Turbine cost:
  A. C. condensing type with exciter
  100 K. W. ......................... $4400.00
  200 K. W. ......................... 6700.00
  300 K. W. ......................... 8000.00
  500 K. W. ......................... 9600.00
  Average price per K. W. ............ 26.00

Switchboard cost about $300.00

Condensers (Jet Type):
  Cost from $1100 for 100 K. W. Turbine up to $2500 for 500 K. W. Turbine.

Cost of power equipment including boilers, turbine, switchboard, condenser, and motors, but not piping, building or smoke-stack, from $80.00 to $90.00 per horse power.

NOTE: These above prices are average costs only and are subject to change with the market.

Relations of the size and speeds of driver and driven pulleys:

$$\text{Diam. of Driver} = \frac{\text{Diam. of driven} \times \text{Rev. of driven}}{\text{Rev. of Driver}}$$

$$\text{Diam. of driven} = \frac{\text{Diam. of driver} \times \text{Rev. of driver}}{\text{Rev. of driven}}$$

# NOTES

# NOTES

## STANDARD APPLIANCES FOR USE IN SHOE FACTORIES

In the past few years the importance of standardized equipment has been demonstrated to a large extent by the demand of manufacturers for indestructible appliances. The metal bench legs, metal racks and other such devices, seem to fulfill this demand most successfully. These metal fixtures become an asset to the manufacturer rather than a liability; for they will wear indefinitely, are practically fireproof, and because of the fact that they are collapsible they are easily taken care of when not in use. These racks, tables, etc., are made up in standard sizes, and, if so desired are made movable with the best type of non-clogging casters or truck wheels; thus making for the greatest convenience.

**Stock Racks** Stock racks are made up in standard size units so that from time to time, as the demands increase, new units may be added and the continuity of the stock room maintained.

**Tables** In most departments of a factory certain hand operations require the use of a table or bench. Much can be said to condemn the old practice of using a continuous bench for these operations, which consume the best light in the less up-to-date factories, and where the amount of bench space required depends largely on the operator's disposition. The individual table takes care of these objections when properly located. Its many uses determining its indispensability, are as follows:

**Cutting Room:** Inspecting, sorting, matching, painting, piecing stays and marking vamps.

**Sole Leather Room:** Inspecting, sorting, doubling, etc.

**Stitching Room:** Cementing, folding, ironing edges, blacking edges, marking for buttonholes, lacing, buttoning, trimming and inspecting.

**Finishing and Packing Room:** Ironing, cleaning, staining, painting bottoms, inking edges, sock lining, repairing tips and vamps, bow attaching, vamp ironing and creasing, lacing and buttoning, inspecting and packing.

**Benches**  Fixed benches are an important fixture in a shoe factory, and a standard metal bench leg has been designed for the convenience of the shoe manufacturer. These are made in two lengths — one for a bench where the operator sits down and the other where he stands up at his work. The expense of these is so slight that the use of wooden joists can hardly ever be considered. Metal legs can be used innumerable times, are always ready for use and make a much neater appearance than wood. The expense of finding a carpenter and looking for material is a decided nuisance as well as a needless expense.

**Mobility of Equipment**  Many operations, such as sorting, stacking soles and counters, etc., have in the past been done on fixed benches or tables, so that the work must be handled several times before it reaches its destination; such as packing and unpacking on trucks. Movable Cut-sole Racks, Movable Sorting Tables, etc., do away with this unnecessary handling by just pushing the table or rack directly to the operator.

Movable Last and Form Racks, Shank Racks, Last Boxes, Upper-carrying Racks, Bag Holders, etc., are helping to carry out this idea of reducing unnecessary handling, which prevents the liability of loss of pieces and the danger of mixing sorted parts, as well as loss of time.

On the following pages will be found illustrations which depict the various types of equipment which have been described in the preceeding paragraphs. They also give an excellent idea of the range of uses in which this equipment has already found a place.

METAL SHOE RACK

(FRAME KNOCKED DOWN FOR SHIPMENT)

NON-CLOGGING CASTOR

LONG DOWEL SHELF

TREEING-ROOM SHELF

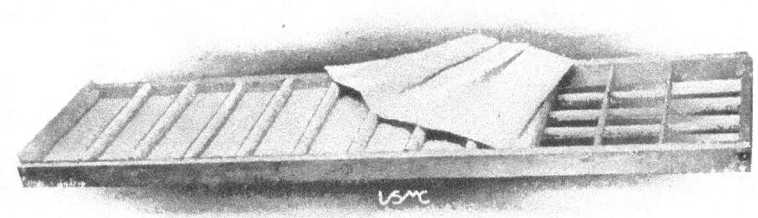

PARTITION SHELF (Long Dowel) WITH COVER

SHOE RACK FITTED WITH COMBINATION PIN PARTITION SHELVES

SHANK RACK

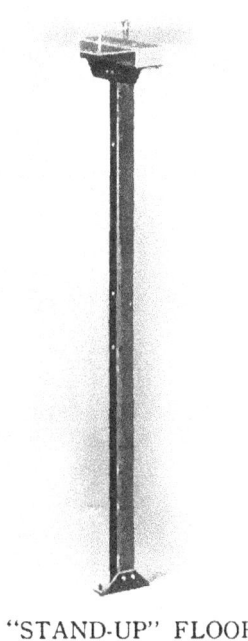

"STAND-UP" FLOOR POST

"STAND-UP" FLOOR LEGS

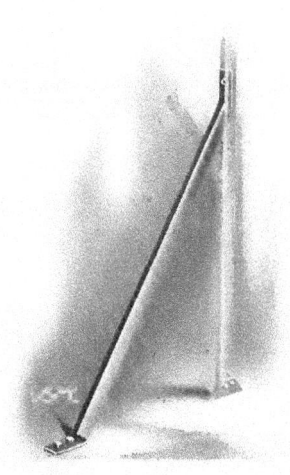

FLOOR BRACE

WALL BRACKET

"SIT-DOWN" FLOOR LEGS

CUT-SOLE RACK

MOVABLE BAG HOLDER

MOVABLE STOCK TABLE

INDIVIDUAL WORK TABLE

UNIT STOCK RACK

UPPER CARRYING RACK

CEMENTED SOLE TRAY RACK

MOVABLE LAST RACK

PACKING AND NAILING RACK

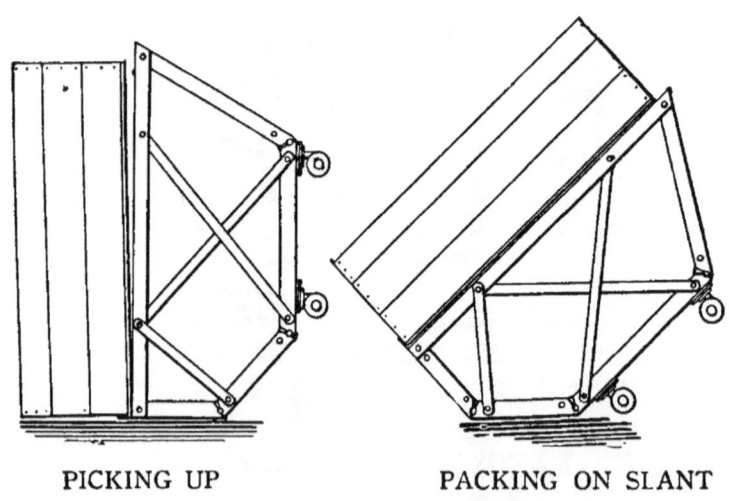

PICKING UP        PACKING ON SLANT

## INDEX

| | |
|---|---|
| Appliances, Standard, for Use in Shoe Factories | 34 |
| Bearings | 23 |
| Bearings, "Sells Roller" | 27 |
| Belt, "Link" Installation in Upper Fitting Room | 24 |
| Belt, Plan of Upper Fitting Tables connected with Spartan "V" — Illustration | 25 |
| Benches | 35 |
| Building, The | 6 |
| Building, The Light of a | 10 |
| Building, Material of | 11 |
| Buidling, Size and Shape of | 8 |
| Cartons, Equipment for Making — Illustration | 28 |
| Carton System, Autobox | 28 |
| Condensers | 22 |
| Dust, Collection of | 24 |
| Endicott-Johnson "Scout" Factory — Illustration | 9 |
| Equipment, Mobility of | 35 |
| Exhaust Fans, Two on Outside Wall — Illustration | 26 |
| Exhaust, Overhead System — Illustration | 25 |
| Figures, Handy Reference | 30 |
| Floors, Altitude of | 10 |
| Floor, Plan, Ideal Shoe Factory | 7 |
| Hospital at Plant of USMC, Beverly — Illustration | 13 |
| Humidifying Room | 14 |
| Humidifying Room — Illustration | 15 |
| Illustrations, Miscellaneous: | |
|     Brace, Floor | 39 |
|     Bracket, Wall | 39 |
|     Castor, Non-Clogging | 36 |
|     Holder, Movable Bag | 40 |
|     Legs, Table, "Sit-Down" | 40 |
|     Legs, Table, "Stand-Up" | 39 |
|     Post, "Stand-Up," Floor | 39 |
|     Rack, Cement Sole Tray | 43 |
|     Rack, Cut Sole | 40 |
|     Rack, Movable Last | 43 |

# INDEX (Continued)

Illustrations, Miscellaneous: (Continued)

| | |
|---|---|
| Rack, Packing and Nailing | 44 |
| Rack, Shank | 38 |
| Rack, Shoe, Combination Pin, Partition Shelves | 38 |
| Rack, Shoe, (Frame knocked down for shipment) | 36 |
| Rack, Shoe, Metal | 36 |
| Rack, Unit Stock | 42 |
| Rack, Upper Carrying | 42 |
| Shelf, Long Dowel | 37 |
| Shelf, Partition (With Cover) | 37 |
| Shelf, Treeing Room | 37 |
| Table, Individual Work | 41 |
| Table, Movable Stock | 41 |
| Lighting | 23 |
| Light and Power, Synopsis | 17 |
| Locker and Wash Room, USMC, Beverly — Illustration | 14 |
| Millwrighting | 23 |
| Millwrighting for upper Fitting Room | 23 |
| Motors, Alternating Current | 20 |
| Motors, High Voltage | 20 |
| Motor, Three H.P., Replacing "Mule Stand Drive" — Illustration | 19 |
| Motor, Five H.P. Driving 12 Hand Method Lasting Machines and Four Rex Pulling-Over Machines — Illustration | 19 |
| Motors, Type of | 20 |
| Offices | 12 |
| Power | 18 |
| Power and Light, Synopsis | 17 |
| Power, Cost of | 23 |
| Power Plant, Cost of | 20 |
| Power, Other Unites | 22 |
| Pulleys | 23 |
| Racks, Stock | 34 |
| Sash, Lupton Steel for Side Walls — Illustration | 11 |

## INDEX (Continued)

| | |
|---|---|
| Sash, Pond Continuous, and Sawtooth Roof — Illustration | 12 |
| Shafting, Speed of | 23 |
| Storage, of Material | 13 |
| Tables, Cutting Room | 34 |
| Tables, Finishing and Packing Room | 34 |
| Tables, Sole Leather Room | 34 |
| Tables, Stitching Room | 34 |
| Tower | 10 |
| Transmission, New Method | 20 |
| Transmission, Old Method | 20 |
| Turbines | 22 |
| Turbine, 100 K.W. Alternating Current Curtis Steam — Illustration | 21 |
| Turbine, 750 K.W. — Illustration | 21 |
| Welfare, Employees' | 13 |
| Windows | 11 |

www.ingramcontent.com/pod-product-compliance
Lightning Source LLC
Chambersburg PA
CBHW022122090426
42743CB00008B/966